Build the Right ~~Fencing~~ for Horses

Jackie Clay

CONTENTS

Introduction 2

Fencing Tools 3

Choosing the Right Fence Posts 8

Anchor Posts and Braces:

    The Backbone of Every Fence...10

Stringing Wire Fencing 17

Putting Up Woven-Wire Fencing 21

The Jackleg Fence..22

Post-and-Rail and Post-and-Plank Fences 23

Fencing the Larger Pasture 26

Fencing the Small Pasture..26

Fencing for Corrals and Paddocks..28

Training Pens 29

The Three Most Common Fencing Problems.....................30

Introduction

Horses are big, strong, playful, and smart, so finding a way to keep them safe and secure in their pastures and paddocks can be a real challenge. Building strong, safe fencing that is also economical and long lasting is an even bigger task. But experienced do-it-yourselfers and even handy beginners can build great fences for their horses.

The considerations for building fencing for horses are quite basic. The fence must be *strong,* because a 1,000-pound (454 kg) horse can exert a lot of power, especially when running, bucking, or playing. However, most fences are broken or ruined not by a fast-moving equine but by a leaning one! Horses are heavy, and the old saying "The grass is always greener on the other side of the fence" seems to have been coined with them in mind. If at all possible, a horse will lean out of a fence, reaching over, under, and through, stretching the wire or even boards to the point where the wire sags, nails pull out, and things break. You must build your fences, then, so that a horse cannot or will not lean over, under, or through them.

The fence must be *safe.* Horses are excitable creatures, sometimes playing rough. A fence needs to be built of material that will keep the animals from being injured while playing, exercising, or just grazing.

A fence built for horses must be strong, safe, and economical.

And for the owner, fencing costs must be considered. Most horse owners would love to have their pasture fenced in on all sides by a 6-foot-high (1.8 m) stone wall — smooth, strong, and extremely safe — but not many of us can afford it! Luckily, there are alternatives, so adequate fencing is available for nearly any budget.

The fence types that best fit these criteria are wire, woven wire, and plank-and-post fences. Your choice will depend on the area to be fenced, how you want it to look, your budget, and the durability you'll require.

Fencing Tools

Tools for horse fencing are relatively inexpensive and basic. Most can be used for several types of fencing and/or other jobs around your farm, yard, or garden, so their cost is minimal for their value. If you are fencing a large acreage, you may choose to hire out to get some of the work done, especially the posthole digging or post pounding. These jobs are quickly done using a tractor with either a posthole auger, which mounts on the rear of a tractor with a three-point hitch, or a post driver, usually mounted on the front of a tractor. A person using a tractor can easily dig more than 100 postholes in part of a day and pound as many fence posts. In contrast, it took me a whole summer to fence a 20-acre (8.1 ha) pasture by hand!

Posthole Digger

If you're opting for the do-it-yourself method of digging postholes for wooden or steel-pipe fence posts, you still have choices. While you can rent or possibly borrow a gasoline-powdered posthole auger, I have had the most luck using an old "armstrong" clamshell digger. The gasoline models work well in moist black earth or fairly moist sand, but they are useless in heavy clays, rocky ground, or hard-packed clay-sand soils. The clamshell digger, on the other hand, will get it all done — with some talent and work. This tool has two cupped steel blades, hinged together, with a handle rising from each. With the handles together, you jab the digger into the earth, usually more than once, to loosen the soil. Then you pull apart the handles while lifting the bite of soil with it. You can also find hand augers, but like the gas type, they're useless on all but excellent soils.

Working with Clay Soil

Digging in clay soil is always a challenge. If the soil is too dry, it must be moistened. (Digging following a mild rain is helpful.) But if the ground is too wet, the clay will stick to the digger. If this is the case, one good remedy is to carry along a small log to rap the clay-clogged digger against. Some folks even resort to dipping the digger in used oil between holes.

As with any other tool, there are a few tricks of the trade in using a posthole digger. Where the soil is dry or compacted, digging is much easier if you dig a shallow, cupped hole through the sod and fill it with water. Then move on to the next few post positions and repeat, allowing time for the water to soak into the soil. You won't believe the difference a little moisture makes, in both the earth and your temper! In very dry conditions, you may have to soak the holes two or three times, but the effort is definitely worth it.

If you must dig holes in very gravelly or rocky ground, using the chisel end of a tamping bar (see page 27) is a great help. Jab it into the hole, prying at different angles to loosen rocks and even to break larger ones. After the rocks and surrounding soil have been loosened, they're usually quite easy to remove with the posthole digger.

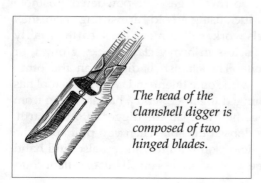

The head of the clamshell digger is composed of two hinged blades.

With some practice, the clamshell digger can be a very effective tool for digging postholes.

Post Driver

If your fence will be made up partly or nearly entirely of steel T-posts (so called because in cross section they are shaped like the letter *T*), you should either buy or borrow a post driver. This is essentially a piece of pipe with a handle on each side and a weighted top end. Slip the driver over the top of the fence post, line up the post, and raise the driver, then bring it down with force. The weight will drive a post much better, and much more safely, than will a sledge-hammer.

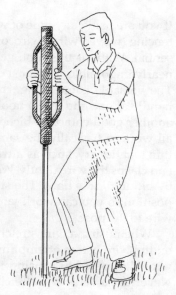

A post driver is the best tool to use for driving metal posts into the ground.

Quick Tips for Using a Post Driver

- If you're having trouble getting the heavy driver over the fence post, try tipping the post first, easing on the driver, and then pulling the entire setup upright. This can really make a difference, especially when you've been working for several hours on a fence line.

- Wear leather gloves and allow your hands to slide on the driver's handles while pounding, instead of forcing it down onto the post. This will help prevent blisters and hand soreness.

- If you are driving posts into very dry soil, soak the area where you will be digging the day before driving to soften the ground. This will make pounding posts a less challenging chore by far!

Wire Stretcher

If you are using any type of wire in your fencing, whether it is stock fencing (woven wire), wire, or even chain link, a wire stretcher is essential. Without this tool, getting the wire tight enough, safely, is nearly impossible. The most common stretcher is a rope-pull model. The versatile come-along, composed of a bar with a ratchet-type handle at one end and a hook with a grasping latch at the other, is another good choice. Whichever kind of stretcher you choose, they all work in basically the same way: One end is attached to something stationary, such as a tractor or a dummy post, and the other end clasps the wire tightly. Work the stretcher to tighten the wire, bit by bit, until it is tight. The stretcher will then hold the wire in that position so you can work safely, with your hands free, to secure the wire to the post.

When using a wire stretcher, be sure it is securely clamped tight on the wire before attempting to stretch it. Otherwise, it can let the wire slip, possibly causing injury to those working on the fence. Never use a vehicle to stretch wire, unless it is as an anchor for the wire stretcher. A vehicle cannot be kept under inch-by-inch control, and the wire can overstretch, snap, and severely injure those on the fence line.

The wire stretcher continues to be useful in the years following the fence building, when you'll use it to take out sags or mend breaks with a solid splice. Best yet, this tool is usually priced very economically!

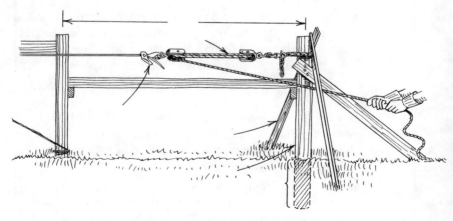

A rope-pull stretcher makes use of a braced dummy post to provide the necessary stability for sufficiently tightening the wire.

Stretching Chain-Link Fencing

If you are using a particularly long run of chain-link fencing, you can usually borrow a chain-link stretcher from the lumberyard or store where you purchased the wire. It differs a bit from a regular wire stretcher but works well to evenly tighten difficult chain link.

Fencing Pliers

When you're working with any type of wire, especially wire that will be stapled to wood posts or fastened to steel T-posts with clips, you need fencing pliers. They can do many jobs — pound or remove staples; fasten clips that bind wire to steel posts; cut, twist, and grip wire; make splices in wire; and more. Also low priced, fencing pliers are a great all-around addition to the toolbox.

Tamping Bar

When wood posts are used, they must be tamped; that is, the earth that fills the hole around the post must be pounded down, bit by bit, until the post is set firmly in the ground. Air pockets and loose soil are the enemy of solid posts. To tamp in a post, you need a tamping bar — a long, heavy rod with a flattened, square end. I've used the handle of a shovel on occasion, but a good steel tamping bar, carrying decent weight, will make the work easier and quicker. As a bonus, most of them have a chisel on the other end that works well for removing rocks from your hole as you dig, breaking rocks, and loosening hard soil.

To properly tamp a posthole, fill in about 4 inches (10 cm) of loose soil, then tamp it firmly by ramming up and down, all around the post, with the tamping bar. Repeat until the hole is overfull, leaving a small mound to settle and ward off rainwater. Do not rock the post to test it, as this will cause air pockets and make the post loose. Leave it alone to set. Also, do not use stones or sod to fill the hole; the stones will often allow air pockets, and the sod may rot over time, causing erosion in the hole and consequently a loose post.

Miscellaneous Fencing Tools

You may also need these additional tools:

- A decent shovel, used mainly to scoop the dirt back into a posthole once the post is in place
- A heavy hammer for pounding staples
- Spikes or "adjusting" braces
- A pair of sturdy leather gloves to protect your hands
- A pair of safety glasses
- A cheerful helper!

Choosing the Right Fence Posts

Three basic types of fence posts are commonly used in building fencing for horses: wood, steel T-posts, and pipe. The smaller the area to be fenced, the stronger the fencing must be. For this reason, our corrals are built using railroad-tie fence posts, planked on top, and heavy-gauge stock panels, while our pasture fences are built simply of wood and steel T-posts and five strands of tight wire.

For most horse owners, the budget is the next most important concern after strength. Take heart! While smooth, geometric, welded pipes are certainly the dream fence for many, fences made from wood fence posts and steel T-posts are not only economical, but when built with care, they are also both durable and attractive.

Wood Posts

There are many different types of wood posts. We like to use railroad ties or used electric poles for corners, braces, and gate posts because they are large in diameter, very strong, and already treated, making them last a long, long time. Your local lumberyard or a neighbor may sell used railroad ties, and the local power company will usually sell used poles at a reasonable price, which can be cut to suit your purpose.

Most folks must use whatever wood fence posts are available locally. Ask around your area to see what other experienced horse owners have used and been satisfied with. Generally, oak, tamarack, juniper, and white cedar make good posts. (In the West, lodgepole pine is often used, but the soil there is usually dry enough that this less-tough wood lasts longer.)

Quick Tip

Do not use poplar, pine, spruce, fir, aspen, or maple as an in-ground fence post — even when treated, these woods will not last long in most soils, making the fence quickly weak and unsafe for the horses.

Lodgepole pine, and sometimes aspen, is used to build *jackleg fencing*, a series of Xs, joined by three or four rails, with no posts set into the ground. The jackleg is an excellent choice of fencing for horses in areas where wood is plentiful — see page 22 for instructions on building your own.

When buying wood fence posts, consider how you will be putting them into the ground. Should you opt to hire a neighbor with a tractor and post pounder — a tool that works very well, and very efficiently, in many areas of the country — buy or cut points on the large ends of the posts before your neighbor comes out. It will save much time, and often it's the only way the posts will drive well. Six-and-a-half-foot (2 m) pointed posts usually work well in most soils. Again, ask the locals!

What diameter posts will you need? This can vary slightly, depending on your horses and what type of posts are available to you, but a 4-inch (10 cm) top with a 6-inch (15 cm) butt (bottom) is adequate for pasture fence posts. For a small pasture, corral, or paddock, I strongly recommend a larger post for greater strength.

Making Your Poles Last Longer

Removing the bark from any wood fence post will make it last years longer, as will treating the at-ground and underground portions of the post. Talk to your County Extension Service agent, usually located in the county seat near you, for recommendations. (We use treated wood — railroad ties and telephone poles — for anchor posts, where durability and strength are of the utmost importance. However, we don't use treated wood for fence posts because of the toxic chemicals it contains, which may leach into the soil — a poor trade for a few more years of use. But a lot of folks do use treated posts. It becomes a personal choice.)

Steel T-Posts

These steel posts are driven into the ground with a post driver; they're quick to use and easy to set, but safest used in pasture fence lines, rather than in corrals, paddocks, or training pens. They have a stabilization plate at their bottom to keep them from rotating once in the ground, as well as small tabs on one side that keep wire strands from slipping up and down the post. A steel T-post is not strong enough to take the stress of a playful horse striking it; it may bend or snap.

Note: Buy the more expensive T-posts, as the cheaper ones are made of poor-quality steel and will easily bend. Six-foot (1.8 m) posts are the best.

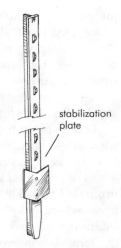

stabilization plate

While not as strong as wooden or pipe posts, steel T-posts are easy to set, making them perfect for fencing larger pastures.

Pipe Posts

Many horse owners prefer, and can afford, to build their entire corral system and pasture fencing out of pipe, welded together as posts and rails. This makes an excellent and very safe horse fence with very low maintenance. In some areas of the country, good used pipe (often from oil-drilling operations) can be had quite reasonably. Look around. If this type of fencing interests you and if someone does have a good pipe fence, drive in and simply ask about his fence. You can usually get considerable information from a satisfied horse person.

Anchor Posts and Braces:
The Backbone of Every Fence

Every fence requires strong corner and gate posts, which are called, collectively, *anchor posts:* stout, deeply set, braced posts that support the corners and gates of your fence. Even welded steel pipe fences benefit from strong corner and gate posts. The corners and gates of jackleg, post-and-rail, and post-and-plank fences are built in a different manner from what's described here, but they, too, are reinforced in their own style — see pages 22 and 23.

Well-built corners, braces, and gates provide the foundation for a good fence. Any wire fence can put an unbelievable amount of strain on the posts over the years. Without exceptional corners and braces, fences soon pull — even those stapled to good solid wood posts. You need only drive through any countryside to see examples of fences with poor corners. You will be able to easily spot them by the leaning corner posts, leaning fence posts, and sagging wire. So take the time necessary to build exceptional corners and braces — even if it takes a day or two per unit — because it will really pay off through the years. Few horses will stay safely contained in a fence that has pulled loose.

Constructing a Corner

When building a fence, we've found it pays big dividends to measure and set all the corners first. I like to use extra-large wood posts (or, as I've mentioned, a used railroad tie or used electric pole, cut down to size) for both the corner and corner brace posts.

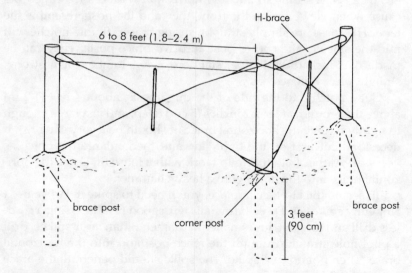

A well-built corner provides stable support for the fence line. To optimize strength, each of the posts should be set at least 3 feet (90 cm) into the ground.

Step 1: Dig holes for and set the three posts.
Each corner is composed of three large, heavy posts: the corner post and two brace posts. The braces should be set 6 to 8 feet (1.8–2.4 m) from the corner post, and all three posts should be set deep — at least 3 feet (90 cm) — into the ground. If a tractor and post pounder are available, and the soil is loose enough so that you can use the pounder on corner posts, it'll make short work of a corner, quickly pounding in the heavy posts and avoiding the need for tamping. Unfortunately, not many folks live in such an ideal location, so the holes must be dug.

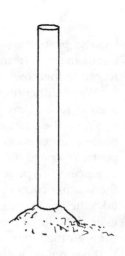

Set the corner and brace posts at least 3 feet (90 cm) into the ground; secure the posts by packing and tamping the earth around them.

Be sure to make the postholes large enough in diameter to allow for adequate tamping of dirt around them. Simply forcing a big post into a small hole will not make a good corner, as there will be air pockets in the soil and later erosion.

Step 2: Construct the H-brace. Once the corner and two brace posts are well set, it's time to add the horizontal H-braces (so called because when viewed from the front, they and the posts resemble the letter *H*). These are simply straight, strong braces, fit into notches cut into the sides of the corner post and two brace posts. They can be made of 2-inch (5 cm) pipe, standard 4 × 4 lumber, or a strong, straight piece of fence post.

Cut a notch into the side of the three posts, about 4 feet (1.2 m) above the ground, at least 2 inches (5 cm) deep, and just large enough to tightly receive the horizontal brace (it should be a very snug fit). If necessary, cut first on the tight side and then enlarge the notch as necessary, rather than trying to work with a loose fit. Drive the horizontal brace into the notch with a heavy hammer.

To secure the H-brace in place, you'll need to spike it. We've used 30-penny spikes and pole-barn nails with good results. Using a cordless drill and long bit, just smaller in diameter than your spikes, drill a pilot hole straight through the fence post and into the horizontal brace. Then pound in a spike. The spike should penetrate the brace by at least 3 inches (7.5 cm). This can also be done when using a 2-inch (5 cm) pipe brace — center the spike into the center of the pipe.

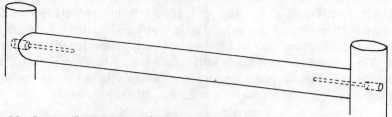

Use long spikes to secure the H-brace to the posts.

Step 3: Wire the H-brace. After the two horizontal braces have been securely fastened in place, it's time to "wire up the Xs." (Yep, I know it's a lot of work, but it does make a tremendous amount of difference in the quality and durability of your fence.) Using any heavy-gauge wire (12 gauge or heavier), measure enough wire to make a loose loop from the top of one brace post to the bottom of the corner post. Cut the wire, then securely twist together the ends. I either cut a slight notch in both posts or drive a staple into place, to permanently hold this loop in position. Remember that you want a loose loop, leaving about 3 inches (7.5 cm) of "slop" in the wire, because you will be twist-tightening it later.

Repeat this process with the other posts, ending up with a wire X between each brace post and the corner. Then hunt up a smooth, straight hardwood stick, about 1½ to 2 inches (3.8–5 cm) in diameter and 18 inches (45 cm) long. A piece of 1-inch (2.5 cm) galvanized pipe will also work well.

Now carefully thread the stick (or pipe) that you're using as a tightener through both loops of the wire, near the center of the X, just below the horizontal brace. Be sure the wire is evenly slack on both sides of the tightener, because if it is not, one wire will end up sloppy. Begin twisting the stick, carefully keeping the wire even and the loop snug. As you twist, you'll notice that

To tighten the wire, thread a stick through both loops of wire; by twisting the stick you will take up the slack in the wires, pulling them taut.

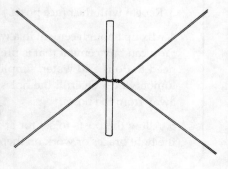

all four wires of the X are becoming snug. When the wires are tight and begin to creak, let the stick go, just past the horizontal brace, and allow the horizontal brace to hold it vertically and securely. If it seems to want to slip, you can secure it in place with a short length of wire.

The X you have just made will hold the post-brace combination tightly together, keep the entire fence from slacking up over time, and effectively make a useful corner, essentially now all one unit.

Corners for Tight Quarters

When your corners are being made for a corral, working pen, or small paddock, it is best to set them into cement.

1. Dig holes for the corner post and braces that are twice the diameter of the posts. To minimize frost heaves, keep the sides of the holes as straight as possible. If frost is a serious problem in your area, flare the bottom of the hole into a bell shape.

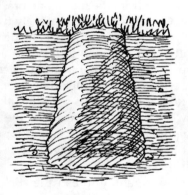

A bell-shaped footing creates a larger mass of concrete, which will be more impervious to the effects of frost.

2. Pour 4 inches (10 cm) of gravel into each hole. Carefully set the post into place, using small, temporary braces as necessary to keep the post upright. Use a level to make sure that it's vertical.

3. Repeat with the brace posts.

4. Mix up enough cement in a wheelbarrow to overfill the holes. (You can buy cement that is premixed with sand so that all you need to do is add water, simplifying the job — ask at the local lumberyard.) Overfill the holes, troweling the cement to slope away from the posts.

5. Allow the posts to set for at least 2 days before you remove the light braces or work on them further.

Gates and Gate Posts

Next you want to make the gates. Gates are made very much like corners, with a sturdy gate post and two H-braces on either side. The gate post itself should be set in concrete. In addition, if the gate is wide and heavy, the gate post should be very long so that you can stretch a guy wire from the post to the gate to keep it from sagging.

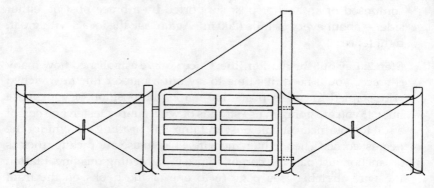

Use H-braces as extra support on either side of a gate.

The best gates available today, other than bolted-together 2 x 6 plank gates, are welded pipe gates. They require little maintenance and are relatively lightweight, very strong, safe for horses, and quite inexpensive. They come in most all widths, from 3-foot (90 cm) walk-through gates to 12- and 14-foot (3.7 to 4.3 m) drive-through gates. They also come complete with hanging hardware, in most cases consisting of a threaded rod, nuts, and washers. Hanging the pipe gate requires a drill and bit of the same size as the threaded rod. Simply drill the holes according to instructions, set the rods in place, and hang the gate level — all said and done, about an hour's work. I have never yet had to replace a pipe gate.

A Gate to Avoid

Don't be fooled into buying a gate made of flat aluminum "lightweight" planks. These are a little cheaper than a pipe gate, and they are quicker and easier to hang. But they twist and break in the wind, are ruined by any strenuous horse activity, and are not worth the time and cost.

H-Brace Supports for the Line

Once your gates are made, it's time to think about the rest of the fence, including the H-braces you'll need in your fence line.

Step 1: Establish how often you need an H-brace. Depending on the soil and the wire you will be using (woven wire requires more bracing than regular wire), you will need an H-brace assembly — composed of an anchor post supported by a brace post on either side — about every ⅛ mile (200 m). Again, ask the locals with great, tight fences.

Step 2: Lay out the marker line. Once you've established how many H-braces you'll need, it's time to lay out a marker line (a stretched wire stapled to corner posts that marks the position of your fence line). If you are going to use strands of wire, simply fasten one end of a roll to your nearest gate post and unroll it, heading straight for the nearest corner. Then tighten the wire. I like to use our pickup truck as an anchor, just past the corner, to use in tightening this line. I fasten the wire stretcher (see page 6) to one of the hooks on the front bumper, or to the trailer hitch, taking care to make it secure, then hook onto the wire. Jack the wire tight, always being careful to stay out of the way — should it break or slip, the wire will spring back and can cut you badly. When the wire seems tight, go back over your line, picking it up gently and allowing it to drop. This will eliminate any bends or bows in your line.

When your line is truly straight, go back and tighten again, if necessary, then staple it tight to the corner post. Very carefully release the wire stretcher. Cut off the wire long enough to securely wrap it around the corner post plus an extra 6 inches (15 cm), but don't wrap it yet — you may need to undo the marker line in order to install the H-braces.

Quick Tip
Even if you do not intend to use wire for your fence, a wire marker line is still handy to mark your fence line. A string or very light wire will not take the stretch necessary to give a straight line.

Step 3: Construct the H-braces. Measure your distances, then make as many H-braces as you need. Assuming that you need to place a brace every ⅛ mile (200 m) — but remember to check with the locals first — a square, 20-acre (8.1 ha) pasture will generally need only one H-brace

between corners. Pastures smaller than that will also benefit from having a supportive H-brace on each side for a total of four — the smaller a pasture, the more "horse activity" will be involved in a shorter length of fence, requiring a stronger, tighter fence than in a big pasture.

Step 4: Erect the H-braces. Erect the necessary H-braces along this wire marker line, and mark the places along the line where the fence posts will be placed to make sure that your fence line is straight and true.

Stringing Wire Fencing

If you want to fence that nice horse pasture with wire, once you have already put in your corners, gates, and any H-braces you need and you have a nice straight wire stretched tightly as a line marker, you're ready to put in fence posts!

Setting Wood Posts

If you've decided on using wood posts, dig a cup through the sod at every 16-foot (4.9 m) interval (at the most) along the marker line. Then dig your holes. It is easiest to remove your marker wire temporarily to facilitate digging. Be sure your fence-post holes are dug at least 2 feet (60 cm) deep. I carry a shovel with a mark burned into the handle at 2 feet, because I've found the more tired I become, or the worse the digging, the more 18 inches looks like 2 feet! I know that if I really do cheat, however, the post will never be solid, and this will weaken the entire fence.

Encounters with Rocks

Should you run into a rock fairly near the surface while digging a hole or driving a post, dig with the chisel end of a tamping bar to either remove the rock or break it. If it is a large, stubborn rock, move your fence post a few inches either way, keeping on the wire line.

Setting T-Posts

If you choose to use steel T-posts, I've found that dropping them from the back of a pickup or flatbed trailer approximately every 16 feet (4.9 m), then going back and measuring each distance, works well. For this chore, I carry a length of heavy twine with a loop at one end to slip over the last post, stretching out 16 feet. It fits in a pocket and makes finding the correct distance easy.

Wear leather gloves to protect your hands, and tip each steel post to slide the pounder on easily. Then line up your post so that the metal tabs found on one of its sides point toward the side your wire will be on. Because you've left your marker wire in place, you can use that to remind yourself where your wire will be, placing the post on the right side of the wire for your purposes. Give a few gentle taps with the driver to set your post and then drive it. It really helps to have a cheerful helper to eyeball the straightness of the post as you are driving. I carry a level to check, after the post is set. Be sure to level it both vertically and horizontally. A straight fence is a thing of beauty that you will be proud of for years to come.

Drive each post well into the ground, 2 or 3 inches (5–7.5 cm) beyond the stabilization plate on the bottom. Very few horses ever jump over a fence. Many more lean out, bit by bit, over the years. The well-set post will not allow this.

Drive each post 2 or 3 inches (5–7.5 cm) beyond its stabilization plate into the ground.

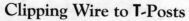

Clipping Wire to T-Posts

When you're fastening wire clips onto T-posts, first slip the clip's short end onto the wire and position the clip around the post. Then grasp the high end of the clip with a pair of pliers and twist the clip around the wire until snug. This will keep the clip from being knocked loose later.

T-post clip

To secure wire to a T-post, fit the clip around the post and then bend the end of the clip to form a hook around the wire.

Stringing the Wire

When all of your posts are in, you can begin stringing wire. It's often best to start with the bottom wire, because this one gets the least use, and you will improve your technique with the practice! Most wire comes on a spool and can be carried easily by two people, each hanging on to the end of a 3- or 4-foot-long (0.9 to 1.2 m), 1-inch-diameter (2.5 cm) pipe stuck through the center of the spool. They must walk evenly or the roll will not unwind well. Each roll of wire will usually do one strand of ¼ mile (400 m) of fence. Wind one end of a roll twice around the anchor post of a gate, then around itself to a length of 6 inches (15 cm). If you're using wooden posts, staple the wire to the post as well. Then set off for the next anchor post (at the next corner or H-brace in the line), unrolling the wire as you go.

At the anchor post, set down the roll of wire. Use a wire stretcher anchored to a stationary object (see page 6) to stretch the wire until it is about as tight as you can jack it. Remember that the wire is tighter at the stretcher than it is in the middle. Then go back, picking the wire up and gently dropping it from time to time, to snap out any bends. Now retighten the wire.

Staple the wire securely to the brace post; if you have used a vehicle as an anchor, staple the wire to the corner post.

Securing wire to wood posts. If you are using wood posts, go back to every other post. Measure up from the ground and staple securely, but don't dent the wire. Marking the wire will weaken it, much the way a wire will break on a slight mark when worked back and forth. After the line is stapled, go back and staple the in-between posts. You'll find that your line will be much straighter this way, even using the measuring stick.

Quick Tip

When stapling, tilt the staple on a diagonal to put the sharp ends in different grains of the wood, making it much stronger and eliminating splitting.

Securing wire to steel T-posts. With steel T-posts, use the above method, but attach the wire to the post with the wire clips that come with the posts and a fencing pliers (see page 19).

After the first wire is secure, if you've used a vehicle as an anchor, carefully release the stretcher. If not, move around the corner and re-stretch the wire, using the corner brace post as an anchor. This keeps the wire very tight. Staple the wire to the corner post, then carefully release the stretcher. In both cases, cut the wire long enough to wrap around the corner post, then tightly around itself for 6 inches (15 cm).

Some folks continue the same wire around a corner, but I don't like to: If anything weakens the tightness of one line of fencing, it cannot weaken the next line as well if it ends at the corner post. I don't know anyone who loves to build fence, or to repair it, myself included, so I try to make my first effort long lasting!

Repeat the above steps with the remaining four or five wires you will be using.

Wire Spacing

I usually carry a long stick marked at 18, 26, 34, 42, and 50 inches (45, 66, 86, 107, and 127 cm). Wires strung at these heights make a good pasture fence for most purposes. If your horses are very tall, or you think they may jump, you can add another wire at 58 inches (147 cm), making a six-wire fence. The reason for this measuring is that few horses will try to reach through such closely spaced wires, eliminating wear on the fence and possible injuries to the horses.

Topping Off

When you're finished stretching the wire, be sure to add a wooden board on the top to prevent horses from leaning over — 1 x 6 hardwood boards or 2 x 6 pine boards work well. Predrilling the boards with a cordless drill that's slightly smaller than your spikes will keep the boards from cracking when you nail them to the posts. Always center the ends of the boards on the fence post for strength. We have had our own cap boards sawed at a local sawmill, from dead trees we cut on our own land. You'd be surprised at how many boards even a medium-sized tree will provide — and how low your cost will be!

Putting Up Woven-Wire Fencing

Most of what I just said about stringing wire fencing can also be applied to woven-wire fencing of all types. There are a couple of differences, though. First, stretching the wire is harder. When you unroll the wire on the ground, try to keep it as straight and tight as possible, and when you stretch it, stretch shorter runs at a time, because it is harder to keep tight. Always stretch on a corner or H-brace, though, or you will loosen posts. And never hook the stretcher to the very end of the fence or to the lightweight vertical wires. It is often easiest to attach a clamp bar or weave a steel pipe through the wire, parallel to the fence post, and use two wire stretchers, one at each end of the bar.

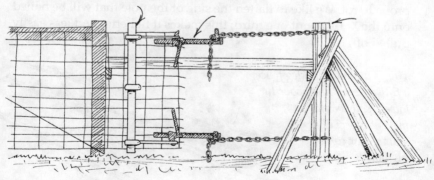

To stretch woven wire fencing, attach two wire stretchers to a bar threaded through the fence.

Once the woven wire is upright against the posts (again, a cheerful helper works miracles here!) and stretched tightly, fasten the heavy horizontal wires with staples or clips, having anchored the wire to an H-brace or corner. Hold the wire 4 inches (10 cm) up from the ground as you go to prevent future rust.

The Jackleg Fence

The jackleg fence is a sturdy rail fence that is easy to put up and can withstand snow loads and wind. Its crossed poles rest on top of the ground, so no postholes are necessary. It is usually about 4½ feet (1.4 m) high and very rustic.

It's easiest to make your first supporting X by using an ax to notch two 5-foot (1.5 m) posts, each about 5 inches (12.5 cm) in diameter, in the high middle and fitting them together. When your first X is just right, lay it on the ground to use as a pattern for the others, which you can fit right in your yard. We spiked the Xs together with two 30-penny spikes, but you can drill and bolt them, if you wish.

When you have a pile of Xs, load them in a truck or on a flatbed trailer and drop one off every 12 feet (3.7 m) or so along the fence line.

The poles should be 13 feet (4 m) long, with several inches of overhang. Peeling the poles with a drawknife is a good idea, as it will make them not only much more eye-appealing but also less prone to rot. We like to flatten the side of the pole that will be nailed onto the Xs by about one third; this makes it fit better and less easily pulled off.

This is a typical jackleg fence; a fifth pole set just above the lowest pole would make it especially suitable for containing horses.

For most horses, three rails below the center of the x is all you need for containment. Place these rails on the inside of the pasture, or the side receiving the most pressure. On our ranch, in Montana, we had to use an additional top pole nestled in the x, not to keep our horses in, but to keep the moose and elk out! Seems like they knew when feeding time was and enjoyed alfalfa hay and sweet feed, too.

Gates can be hung by digging a good posthole snug next to an x, tamping in a large, tall anchor pole, and spiking the post into the intersection of an x. There will be a similar post on the other side of the gate opening to fasten the other end of the gate to.

Post-and-Rail and Post-and-Plank Fences

A wood post-and-rail or post-and-plank fence is beautiful, safe, and long lasting. Use care to select the best material you can get for the rails or planks. Ask some local horse owners with fences that you've envied for suggestions. Folks love to talk about accomplishments that they're justly proud of. Generally, materials such as poplar, aspen, jack pine, and spruce make poor rails or fence planks. They rot too quickly and are seldom strong enough to work satisfactorily. Hardwood usually makes good fencing material, but you should predrill each rail to prevent splitting and nail bending. Likewise, tamarack makes good rails; it's tough, long lasting, and straight.

Like wire fences, post-and-rail and post-and-plank fences need strong, sturdy corner posts and gate posts. You may want to consider anchoring these posts in a cement base. These posts do not, however, need to be further supported by H-braces.

Building the Post-and-Rail Fence

Rails for this fence should generally be at least 3 inches (7.5 cm) thick, but not more than 5 inches (12.5 cm). Thinner rails will break and sag, while thicker rails will be very heavy and hard to nail unless you use bolts or lag screws. In addition, rails should extend for no more than 12 feet (3.7 m) between posts or they will sag.

It is best to flatten the ends of the rails by about a third for ease of nailing and strength of the fence. This can be done with a chain saw, an ax, or a drawknife.

There should be four rails on a small pasture fence, with the top rail about 4½ feet (1.4 m) off the ground, followed by rails at 3½ feet (1 m), 2½ feet (75 cm), and 12 to 18 inches (30 to 45 cm) from the ground. You can adjust the measurements to fit your horses if they differ from the norm — if they're saddlebreds, for instance, or hunters, draft horses, or even miniature horses.

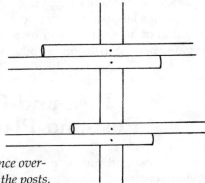

The rails on a post-and-rail fence over-lap, one on top of the other, at the posts.

The Pros and Cons of Electric Fencing

I haven't included electric fencing in this discussion, for several reasons. First, only the very expensive electric fences, such as the high-tensile New Zealand fence and the woven poly-mesh fence, will adequately contain horses. All too often, I've seen horse pastures made up of one or two strands of light electric wire strung between far-apart lightweight posts. These will sometimes work for divisions in a pasture, allowing for grazing rotation, but horses, in playing or when spooked, run into this light wire, get shocked, and usually end up leaping right through the fence. Any loose horse is dangerous, especially if you live on a well-traveled road.

In addition, problems such as fire caused by a downed wire, shorted-out fencing due to blowing weeds, trees falling on the wire, and deer breaking the fence quickly diminish the allure of building an electric fence around a horse pasture.

Building the Post-and-Plank Fence

Like the post-and-rail fence, a post-and-plank fence is safe and sturdy, with the added benefit of being very beautiful. With the posts secured, boards or planks take the place of rails. Although 1-inch (2.5 cm) rough-sawed hardwood boards are often used successfully in horse fencing, it is much better to use 2 x 6 planks, unless the fence posts are moved in and spaced only 10 feet (3 m) apart. The planks should be sawed individually, measured to meet in the center of the post. Even 2-inch (5 cm) pine or fir planks are best predrilled before being spiked onto the post, to prevent splitting of the ends. Any splitting weakens the fence.

Any wood post-and-rail or -plank fence lasts longer if it is treated with weather sealant or

> ### Quick Tip
>
> Some folks further strengthen a post-and-plank fence by adding a 2-foot (60 cm) piece of 2 x 6 over the post-plank joint and nailing, bolting, or lag-screwing this reinforcement to both the fence planks and the post.

oil stain. This is usually better than painting — unless you have help in keeping that horse-pasture fence glistening white, which means painting, painting, and more painting.

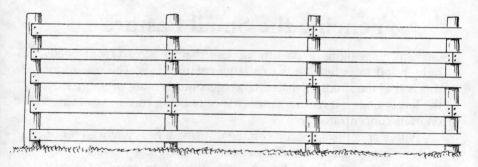

Posts and planks make a beautiful, traditional fence.

Fencing the Larger Pasture

Fencing a large pasture often appears to be an overwhelming job. Tackled in a systematic way, however, it can soon be cut down to a manageable size. First, you must decide what kind of fence and fence posts you want (or can afford). With a larger pasture, there are many choices, depending on economics and location. For instance, the cheapest and easiest fence to construct for a large pasture is a tight, five-strand, 12-gauge wire fence, clipped securely onto steel T-posts. With this fence, only the corners, gate braces, and H-braces need to be wood posts (which require the work of hand- or tractor-dug holes and tamping by hand). Therefore, the fence goes up quite quickly (especially with some help) and is inexpensive.

If you can afford it, heavy-gauge woven stock fencing with either a hardwood 1-inch (2.5 cm) or a pine 2-inch (5 cm) board on top of it makes a good pasture fence. The top board keeps horses from leaning over for that greener grass on the other side.

Another option is a post-and-rail fence, using wood posts spaced 12 feet (3.7 m) apart and either rails or 2-inch (5 cm) planks nailed on 12 to 15 inches (30–37.5 cm) apart. Alternatively, a jackleg fence can use the same dimensions, but requires no posts in the ground. Any of these makes an excellent, safe horse-pasture fence.

Fencing the Small Pasture

Generally speaking, a pasture of less than 5 acres (2 ha) needs tighter and safer fencing than a large pasture does. The reason for this is that the smaller pasture fence gets much more pressure than the larger one — it has a greater chance of being run into or leaned on in its concentrated area.

With a small pasture, I recommend using a heavy woven fence, such as a triangle "coyote" fence or woven 2 x 4 horse fencing, topped with a sturdy cap board or a heavy-gauge strand of electric fencing, a jackleg fence, or a wood post-and-rail or -plank fence.

The basics of wire fencing for small pastures are about the same as for large, but I opt for spacing the posts 12 feet (3.7 m) apart to concentrate the strength of the fencing. The more a fence is apt to get abused, the closer together its posts must be; it also needs to be

A post-and-plank fence is sturdy enough to use in tighter spaces, such as small pastures, corrals, and paddocks.

stronger, safer, and higher. This is why stallion runs are often 6 feet (1.8 m) high, built of heavy chain link, 2 x 6 planks, or some such tough material: secure, strong, and safe.

As for the options in wood fencing, much will depend on what area of the country you live in, unless your finances allow freedom of choice. In the western parts of North America, especially near the mountains, a jackleg fence makes an excellent long-lasting option for a small pasture. Frequently built from standing dead lodgepole pine or stands of lodgepole that need thinning, these are inexpensive (although a bit labor intensive) and absolutely beautiful. Posts and poles may frequently be had inexpensively by obtaining a "pole patch" from your local forest service. You will pay a small fee, be assigned an area to selectively cut (which thins stands for further growth), and be able to cut trees from which to make your fence.

Fencing for Corrals and Paddocks

Because corrals and small paddocks are fairly tightly enclosed spaces, receiving much pressure from horses, they must be very securely built to be safe and lasting. Whereas in the large pasture fence posts could be smaller and 16 feet (4.9 m) apart, in a corral or lot wood posts need to be a minimum of 8 inches (20 cm) in diameter and set no farther apart than 10 feet (3 m). Our ranch uses railroad ties every 10 feet, with a steel T-post in between and heavy-gauge welded stock panels on the inside, capped by 2 x 8 planks to which the T-posts are securely fastened, as are the railroad ties. This has worked very well, with no damage to the fence or horses.

Welded steel pipes with pipe fence posts set into concrete also make a very good small-area fence, but unless you have an economical source of material and can weld, the cost may become prohibitive.

A wood post-and-rail or post-and-plank fence also makes a very good corral or paddock choice. Again, though, in any heavy-traffic or high-pressure area it should be stronger, tighter, and higher than a pasture fence. Use heavier, larger fence posts, rails that are larger in diameter, and planks at least 2 x 6; set the rails or planks closer together, too, adding two more to each section of fencing.

A heavy woven-wire fence, such as a triangle fence (often called coyote-proof or non-climbable), or woven 2 x 4 horse fencing, may be incorporated into a corral or small-lot fence, but it should be reinforced with rails or planking to keep the horses away. If this is not done, then sooner or later the wire will begin to bag and sag from constant contact, making the fence deteriorate quickly. Likewise, chain-link fencing 6 feet (1.8 m) high is often used for horse pens. But I still prefer to see it reinforced with wood or pipe, because like the woven wire, it can develop bags from being kicked, leaned on, or rubbed. (In real life, more fences are ruined by itchy necks and backsides than by wild horses!)

A good alternative to woven wire in corral building is heavy welded stock panels. These come in "cattle," "bull," and "hog" heights. The hog panels are too short for corral use. The bull panels are 5 feet (1.5 m) high, but are less often available. Cattle panels, which range from 48 to 56 inches high, are easily available through most lumberyards and farm stores, and work well for many applications. Avoid them, however, if you're containing very active or many young horses; these could kick through the 6-inch by 6-inch (15 x 15 cm) mesh and possibly be injured. The stock panels can be easily stapled to the inside of railroad ties or solid wood posts, or welded to steel posts set into concrete. They should be capped with either a pipe or a plank to prevent damage to the fence or to a horse.

Training Pens

Outside of a stall, the training pen is the tightest area to be fenced. Taking into consideration the amount of intense use a training pen gets through the years, its fence should be very tight. Most horse owners like a training pen that is at least 6 feet (1.8 m) high (although some prefer 8 feet, or 2.4 m) and very solidly built. Many folks like a wood post-and-plank fence with plywood enclosing the pen, like a room — all solid, non-see-through walls. I prefer a more open, but still stoutly built, post-and-plank fence. Like most things, this is a personal decision. But the one thing all training pens, working chutes, and horse restraining areas should have in common is strength and safety for horse and owner. Barbed wire has absolutely no place in or near such an area, nor do steel T-posts. (I've seen both horses and trainers injured on steel posts.) Much better to have a solid, safe training ring!

Combination Fences

Often your best choice is a combination of fencing for horses. Areas that receive a lot of use and stress, such as feeding, watering, and training areas, benefit from extra-tight fencing — even if this means fencing certain sections of a large pasture with wire. Some horse owners reinforce a simple fence with several sections of stock panels, wood or pipe rails, or planks. This is an excellent idea that protects both the fence and the horses.

Any place where the pasture becomes smaller and crowds the horses is a good candidate for reinforcement. Gates, corners of smaller pastures, areas near the barn, and areas where mares may gather to "visit" a stallion in a lot across the way are all potential problem spots.

The Three Most Common Fencing Problems

Well, here you are, building fencing for the horse pasture. Suddenly you discover that a section of the fence, or a corner, ends up in a swamp or on a solid rock hillside! (We found out that a corner of our fence in Montana was on a rock ledge for 50 feet, or 15 m, on either side.) Now what? The most common problems that fence builders encounter have to do with too much rock and too much water. Here are the solutions that have worked for us.

Q My fencing line runs across a rock ledge. How do I run my fence line across it if I can't drive the poles and posts in?

A When you find yourself fencing on solid rock, adding a section of jackleg fence works great. Find the last "good" place for your regular fence, then dig in an H-brace, if you're a ways off from a set corner or brace. Simply add the jackleg sections to and away from your corner, then continue on your way, adding another H- brace as soon as you are able to dig into the ground.

Q There's a huge rock in the ground where my anchor post is supposed to go. What should I do?

A When you discover a big rock, or several big rocks, where you must dig a corner, the best option is to enlarge the hole and remove them. You can do this by using the chisel end of a tamping bar, by driving a heavy steel pry bar into the ground, or by digging with the shovel and posthole digger. When you finish, it may look like you were mining instead of fencing, so be sure you tamp in that corner post especially well.

Q My fence line runs across a swampy area. What's the best way to run my fence across it without compromising the fence's strength and stability?

A Swampy areas present a different problem. The posts go in too easily and just about disappear with a few solid blows from a post pounder. For fencing swampy areas, it is often a good idea to use pointed wood posts; drive them with a maul while you stand on a solid sawhorse or specially built box. The wood posts will hold better than a steel T-post in a wet area. Be sure to use either treated posts or wood that performs well in damp applications, such as cedar, cypress, or tamarack.

It is nearly impossible to dig postholes in swampy areas, and even worse to tamp them in solid. Driving becomes the only viable alternative. If you are able to use a tractor and post driver, do so; this method allows you to drive longer, and thus more solid, posts. Working in the dry part of the summer also offers you a better chance of success, especially if you're using heavy equipment.

Other Storey Books You May Enjoy

101 Horsemanship & Equitation Patterns: A Ringside Guide for Practice and Show, by Cherry Hill. Step-by-step, illustrated instructions for the most widely used patterns in the most popular classes of Western and English competition. The book's convenient vertical, comb-bound format allows readers to hang it in the barn or lay it on a barrel for reference. Includes a full-page arena map for every pattern. Paperback with comb binding. 224 pages. ISBN 1-58017-159-1.

Fences for Pasture & Garden, by Gail Damerow. The complete guide to choosing, planning, and building today's best fences: wire, rail, electric, high-tension, temporary, woven, and snow, plus chapters on gates and trellises. Paperback. 160 pages. ISBN 0-88266-753-X.

Horse Handling & Grooming: A Step-by-Step Photographic Guide, by Cherry Hill. This user-friendly guide to essential skills includes feeding, haltering, tying, grooming, clipping, bathing, braiding, and blanketing. Practical advice is offered throughout for both beginners and experienced riders. Paperback. 160 pages. ISBN 0-88266-956-7.

Horsekeeping on Small Acreage: Facilities Design & Management, by Cherry Hill. The essentials for designing safe and functional facilities whether you have one acre or hundreds. Paperback. 192 pages. ISBN 0-88266-596-0.

Making Your Small Farm Profitable, by Ron Macher. This practical, step-by-step guide to operating a small farm in the new millennium examines 20 alternative farming enterprises. Readers will learn how to target niche markets and sustain a farm's biological and economic health. Paperback. 288 pages. ISBN 1-58017-161-3.

Starting & Running Your Own Horse Business, by Mary Ashby McDonald. This creative guide shows readers how to run a successful business and make the most of their investments in horses, facilities, and equipment. Includes dozens of essential forms and contracts. Paperback. 160 pages. ISBN 0-88266-960-5.